Pasc
D0071111

Popular Lectures in Mathematics

Survey of Recent East European Mathematical Literature

A project conducted by
IZAAK WIRSZUP,
Department of Mathematics,
the University of Chicago,
under a grant from the
National Science Foundation

V. A. Uspenskii

Pascal's Triangle

Translated and adapted from the Russian by David J. Sookne and Timothy McLarnan

The University of Chicago Press
Chicago and London

The University of Chicago Press, Chicago 60637
The University of Chicago Press, Ltd., London

Printed in the United States of America

International Standard Book Number: 0–226–84316–5
Library of Congress Catalog Card Number: 73–90941

V. A. USPENSKII, a well-known mathematical logician, is a professor at Moscow University. [1974]

Contents

Preface

The reader who is not familiar with Pascal's triangle should be warned that it is not a geometric triangle with three angles and three sides. What we call Pascal's triangle is an important numerical table, with the help of which a number of computation problems may be solved. We shall examine some of these problems and shall incidentally touch upon the question of what "solving a problem" can mean in general.

This exposition requires no preliminary knowledge beyond the limits of the eighth-grade curriculum, except for the definition of and notation for the zeroth power of a number. That is, one must know that any non-zero number, raised to the zeroth power, is considered (by definition!) to be equal to unity: $a^0 = 1$ for $a \neq 0$.

1 A Problem from the Eighth Olympiad

During the Eighth Moscow Mathematical Olympiad (1945), the following problem was presented to the ninth- and tenth-grade participants:[1]

A network of roads is given (fig. 1.1). From point A, 2^{1000} men set out. Half go in direction l, half in direction m. Reaching the first intersection, each group divides; half go in direction l, and half in direction m. Such a division takes place at each intersection. How many people arrive at each intersection of the 1000th row?[2]

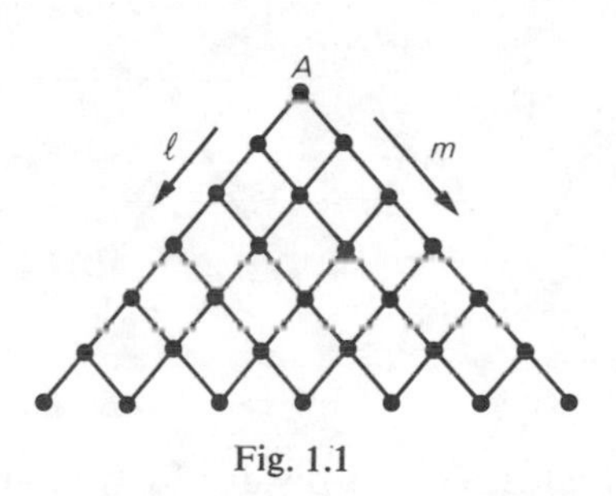

Fig. 1.1

First let us observe that at the moment, we do not know whether the problem has a solution; that is, whether the division of people can proceed as required by the problem's conditions. We know that if an odd number of people arrive at some intersection at which the usual division of the stream of people is to take place, then the division is blocked. Consequently, for the problem to have a solution, it is necessary and sufficient that an even number of people arrive at each intersection of each of the first thousand rows, from the zeroth to the nine hundred ninety-ninth. We must make certain that this is so in the process of solving the problem.

Let us begin by introducing symbols for the number of people who pass

1. See A. M. Yaglom and I. M. Yaglom, *Challenging Mathematical Problems with Elementary Solutions* (San Francisco: Holden-Day, 1964), 1:19, problem 62b.

2. Consider the rows to be numbered, starting with the zeroth. Thus, in the zeroth row, there is one intersection (A); in the first, two; in the second, three; and so on.

through each intersection of our network of roads. The intersections of each row will be numbered from left to right beginning with the zeroth; consequently, the intersections of the nth row will be numbered from zero to n. The number of people who pass through the kth intersection of the nth row will be denoted by $H^n{}_k$. Since it is not clear at the present time that the problem has a solution, we cannot be certain that *all* numbers $H^n{}_k$ exist; that is, the number $H^n{}_k$ exists for each k from 0 to n and for every n from 0 to 1000. It is, however, clear that some of these numbers exist. By virtue of the notation we have introduced,

$$H^0{}_0 = 2^{1000}. \tag{1.1}$$

Let us now determine how the numbers $H^n{}_k$ ($k = 0, 1, 2, \ldots, n$) and $H^{n+1}{}_k$ ($k = 0, 1, 2, \ldots, n + 1$) are related, under the supposition that they all exist. We shall show that if all the numbers $H^n{}_k$ exist and are even, then all the numbers $H^{n+1}{}_k$ exist. Let us examine the nth and $(n + 1)$th rows of intersections and the road segments which connect them. At each intersection we place the appropriate symbol for the number of people arriving (see fig. 1.2). The number of people who enter the

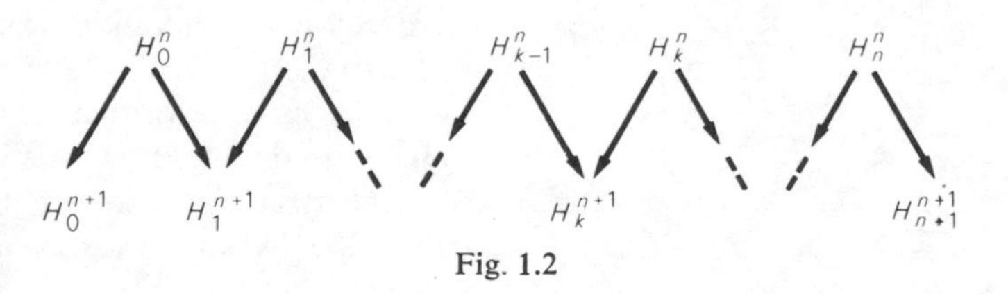

Fig. 1.2

zeroth intersection of the nth row (that is, $H^n{}_0$) is divided by two, and only half of these people enter the zeroth intersection of the $(n + 1)$st row; therefore,

$$H^{n+1}{}_0 = \frac{H^n{}_0}{2}. \tag{1.2}$$

The other half of the $H^n{}_0$ people enter the first intersection of the $(n + 1)$th row and there join half the people who left the first intersection of the nth row, who number $H^n{}_1/2$.

Therefore, $H^{n+1}{}_1 = (H^n{}_0 + H^n{}_1)/2$. In general, the number of people who enter the kth intersection of the $(n + 1)$th row is the sum of half the number of people who left the $(k - 1)$th intersection of the nth row (or $H^n{}_{k-1}/2$), and half the number of people who left the kth intersection of the nth row (or $H^n{}_k/2$). Thus,

$$H^{n+1}{}_k = \frac{H^n{}_{k-1} + H^n{}_k}{2} \quad \text{for } 1 \leq k \leq n\,. \tag{1.3}$$

Finally, the number of people who enter the $(n + 1)$th intersection of the $(n + 1)$th row is equal to half the number of people who left the nth intersection of the nth row:

$$H^{n+1}{}_{n+1} = \frac{H^n{}_n}{2}\,. \tag{1.4}$$

The relations (1.1)–(1.4) allow us to establish the fact that the problem has a solution. Actually, from equations (1.2)–(1.4) it follows that if for any fixed n all numbers of the nth row ($H^n{}_0, H^n{}_1, \ldots, H^n{}_n$) exist and are divisible by $2a$, then all numbers of the $(n + 1)$th row ($H^{n+1}{}_0, H^{n+1}{}_1, \ldots, H^{n+1}{}_{n+1}$) exist and are divisible by a. For if we suppose that $H^n{}_0, H^n{}_1, \ldots, H^n{}_n$ exist and are all divisible by $2a$, then there are integers (whole numbers) $M^n{}_0, M^n{}_1, \ldots, M^n{}_n$ satisfying the relations

$$H^n{}_0 = 2aM^n{}_0$$

$$H^n{}_1 = 2aM^n{}_1$$

$$\cdots$$

$$H^n{}_n = 2aM^n{}_n\,.$$

Thus, we have (using (1.2)–(1.4)):

$$H^{n+1}{}_0 = \frac{H^n{}_0}{2} = aM^n{}_0\,;$$

$$H^{n+1}{}_k = \frac{H^n{}_{k-1} + H^n{}_k}{2} = \frac{2aM^n{}_{k-1} + 2aM^n{}_k}{2}$$

$$= a(M^n{}_{k-1} + M^n{}_k) \quad \text{for } 1 \leq k \leq n\,;$$

$$H^{n+1}{}_{n+1} = \frac{H^n{}_n}{2} = aM^n{}_n\,.$$

This establishes the claim that the numbers $H^{n+1}{}_0, H^{n+1}{}_1, \ldots, H^{n+1}{}_{n+1}$ exist and are all divisible by a.

Therefore, since all numbers of the zeroth row (there is only one, $H^0{}_0$) exist and are divisible by 2^{1000} (by 1.1), we have verified that all numbers of the first row,

$$H^0{}_1, H^1{}_1,$$

exist and are divisible by 2^{999}; all numbers of the second row,

$$H^2{}_0, H^2{}_1, H^2{}_2,$$

exist and are divisible by 2^{998}; and so on, until all numbers of the 999th row,

$$H^{999}{}_0, H^{999}{}_1, \ldots, H^{999}{}_{999},$$

exist and are divisible by 2; and all numbers of the 1000th row,

$$H^{1000}{}_0, H^{1000}{}_1, \ldots, H^{1000}{}_{1000},$$

exist (and are divisible by 1).

The relations (1.2)–(1.4) not only show that the problem has a solution, but also provide a method for calculating the line of numbers

$$H^{n+1}{}_0, H^{n+1}{}_1, \ldots, H^{n+1}{}_{n+1}$$

from the line

$$H^n{}_0, H^n{}_1, \ldots, H^n{}_n.$$

Repeatedly applying these relations, beginning with the zeroth line (by using (1.1)), we theoretically can calculate the numbers $H^n{}_k$ for all 501,501 intersections in all the rows through the 1000th and, in particular, for all intersections of the 1000th row, thus solving the problem. The direct calculations for the first rows are:

$$H^1{}_0 = \frac{H^0{}_0}{2} = \frac{2^{1000}}{2} = 2^{999}; \quad H^1{}_1 = \frac{H^0{}_0}{2} = \frac{2^{1000}}{2} = 2^{999};$$

$$H^2{}_0 = \frac{H^1{}_0}{2} = \frac{2^{999}}{2} = 2^{998}; \quad H^2{}_1 = \frac{H^1{}_0 + H^1{}_1}{2} = \frac{2^{999} + 2^{999}}{2} = 2^{999};$$

$$H^2{}_2 = \frac{H^1{}_1}{2} = \frac{2^{999}}{2} = 2^{998}; \quad H^3{}_0 = \frac{H^2{}_0}{2} = \frac{2^{998}}{2} = 2^{997};$$

$$H^3{}_1 = \frac{H^2{}_0 + H^2{}_1}{2} = \frac{2^{998} + 2^{999}}{2} = 3 \cdot 2^{997}; \text{ and so on}.$$

2 What It Means to Solve a Problem

Thus, the problem of chapter 1 is solved . . .

"How is it solved?" wonders an unconvinced reader (the convinced reader knows in advance what the author is going to say, and nothing makes him wonder). "I don't see that we have solved it."

AUTHOR: Well, of course we have solved it. You know that to solve a problem means to find its solution. And we have just found the solution.

READER (*indignantly*): Is this really a solution?

AUTHOR (*pretending that he doesn't understand what the trouble is*): Well, is it really incorrect?

READER: No, "it" is correct, but "it" is not a solution.

AUTHOR: But then what is a solution?

READER: A line of numbers showing how many people arrive at each intersection of the thousandth row.

AUTHOR: But there would be 1001 numbers in this line. Is it possible that the organizers of the Eighth Olympiad wanted the participants to write 1001 numbers?

READER *becomes thoughtful.*

AUTHOR: I have a proposal. So as not to complicate the situation with long sequences, let us select one intersection, and concern ourselves with the number of people who arrive there. All right?

READER *agrees.*

AUTHOR: Now what will we consider to be a solution to the following problem: How many people arrive at the third intersection of the fourth row?

READER: Huh? A number.

AUTHOR: Written how?

READER (*amazed*): Well, in the decimal system.

AUTHOR: But isn't an answer like "$H^4{}_3$" a solution?

READER: Of course not. Some solution!

AUTHOR: By continuing the series of calculations which we began at the end of the preceding section, it is easy to verify that 2^{998} people visit the third intersection of the fourth row. Will the answer "2^{998}" be a solution to the problem?

READER (*still not seeing the trap*): Yes, of course.

AUTHOR: But you know that the expression "2^{998}" is not an expression in the decimal system. This expression consists of *two* decimal numbers, "2" and "998," whose relative position shows what operation must be performed upon them in order to obtain the desired number.

READER: But the expression "2^{998}" can easily be converted into decimal form.

AUTHOR: Not so easily; just try to raise 2 to the 998th power. But that isn't even the trouble; the trouble is that just now you contradicted your previous statement. Earlier, you agreed to consider only a number written in the decimal system as a solution. From the point of view of this definition, the expression "2^{998}" is still not a solution (it is a so-called half-finished product from which the solution may be derived). Of course, such a point of view is acceptable only if it is held consistently. But another point of view is possible, according to which 2^{998} is a solution. Such a point of view will probably clarify the matter for you. You know that often the simplest answers to mathematical problems come not directly in the form of a number written down in the decimal system, but in related "indirect" form. With this in mind, what should we settle for as a "solution," in our example, to the problem of how many people visit the third intersection of the fourth row?

READER: In our example, we must accept as a solution any expression for which there is a method which lets us get a numerical answer (written in the decimal system) from that expression. That is, 2^{998} will be a solution. Although the method of getting a decimal answer (997 consecutive multiplications) is long, it is feasible in principle.

AUTHOR: But then why isn't $H^4{}_3$ a solution? Here, too, there is a method of getting a decimal answer. It is given by the relations (1.1)–(1.4).

READER is perplexed.

AUTHOR (*satisfied that he has succeeded in leading the reader into a blind alley—the inexperienced reader, that is: the experienced reader will himself lead the author into a blind alley*): The point is that there are at least three interpretations of what we mean by a solution to the problem of the number of people who visit a given intersection.

First interpretation: By a solution, we mean a number written in the decimal system.

Second interpretation: By a solution, we mean some expression which designates a number, and for which a method is known that allows us to get the designated decimal number (a so-called first-interpretation solution) from the expression.

Third interpretation: By a solution, we mean some expression which designates a number (written in the decimal system) and which is made up of numbers and some operations considered "standard" (for instance, the usual arithmetic operations).[1] We require that each standard operation be accompanied by a method of getting the decimal result from the decimal numbers to which it is applied (as is the case with the arithmetic operations). Then for each expression which is allowed, there will exist a method allowing us to get the decimal number designated from the decimal numbers which are part of the expression, so that a solution under the third interpretation will automatically be a solution under the second.

Under the first interpretation, neither H^4_3 nor 2^{998} is a solution to the problem of finding the number of people arriving at the third intersection of the fourth row. To get a solution, we must find a decimal expression for 2^{998}; however, this expression would consist of more than 300 digits, and could be calculated in a reasonable amount of time only by a computer.

Under the second interpretation, both H^4_3 and 2^{998} are solutions.

In the case of the third interpretation, everything depends upon the choice of the initial standard operations: If exponentiation is included among them, then 2^{998} will be a solution; if it is not included, then 2^{998} will not be a solution. If the standard operations include the operation H which calculates the number H^n_k from the numbers n and k (note that the relations (1.1)–(1.4) give a method of performing such a calculation, so that the requirement we imposed on standard operations is satisfied), then H^4_3 will be a solution to the problem; in the opposite case, it will not.

The question of whether we may choose the standard operations arbitrarily naturally arises. Speaking formally, we may certainly do so. In practice, of course, we should choose as standard operations (through which we are required to express the solution of any problem) such operations as are encountered in the solutions of many problems, or at least in the solutions of important ones. Such operations include the four arithmetic operations and several other operations, such as

1. The set of standard operations must be indicated beforehand. It is important to emphasize that the third interpretation depends on the choice of this set. Thus, the expression 2^{998} will be a solution under the third interpretation precisely if the operation of exponentiation is included as one of the standard operations.

exponentiation and the operation of taking factorials (see below, chapter 6). If the operation H were needed for important problems or if our own problem about intersections were very important, then perhaps the operation H would deserve to be ranked with the standard operations. However, the operation H was undeserving before we introduced our problem, and is scarcely worthy now. In section 4 we will examine an operation similar to H which, as we shall see, deserves to be included as a standard operation.

But now we must return to our original problem of the intersections of the thousandth row. Its solution may be sought in three forms, corresponding to the three interpretations of a "solution" described above.

1. In the form of a sequence of 1001 numbers, written in the decimal system. We shall not seek such a solution (since we found it too difficult even to find such a solution for one intersection of the fourth row).

2. In the form of an expression which in principle allows us to calculate the number (that is, to find a decimal representation of the number) of people who arrive at each intersection of the thousandth row. We have already found such a solution: $H^{1000}{}_k$, for which the process of calculation is given by the relations (1.1)–(1.4).

3. In the form of an expression which not only allows us to calculate $H^{1000}{}_k$ for any k from 0 to 1000, but which is formed by means of certain standard operations. It is in this form that we shall seek a solution. In the following exposition it will become clear precisely which operations will be considered standard for our purposes.

3 Pascal's Triangle

Let us consider a sequence of numbers $d_0, d_1, \ldots, d_n$, for some $n = 0, 1, 2, \ldots$ (for $n = 0$ this sequence "degenerates" to the sequence consisting of the single number d_0). From it, let us generate a new sequence of numbers $s_0, s_1, \ldots, s_{n+1}$, by the following rule:

$$s_0 = d_0\,, \tag{3.1}$$

$$s_k = d_{k-1} + d_k \qquad (1 \leq k \leq n)\,, \tag{3.2}$$

$$s_{n+1} = d_n\,. \tag{3.3}$$

We say that this new sequence is derived from the original one by *Pascal's relations*. For example, from the sequence 2, 0, 2, we may produce (using Pascal's relations) the new sequence 2, 2, −2, −2, and from this one in turn the sequence 2, 4, 0, −4, −2.

The French mathematician and philosopher Blaise Pascal (1623–62) investigated the properties of a triangular table of numbers, each row of which is derived from the preceding by the relations (3.1)–(3.3). This table, which we shall examine further, is now known as "Pascal's triangle."

Remark 1. If the sequence β is derived from the sequence α by Pascal's relations, then the sum of the elements of sequence β is equal to twice the sum of the elements of sequence α. For using the relations (3.1)–(3.3),

$$\begin{aligned} s_0 + s_1 + s_2 + \cdots + s_n + s_{n+1} \\ &= d_0 + (d_0 + d_1) + (d_1 + d_2) + \cdots + (d_{n-1} + d_n) + d_n \\ &= (d_0 + d_0) + (d_1 + d_1) + \cdots + (d_n + d_n) \\ &= 2(d_0 + d_1 + \cdots + d_n)\,. \end{aligned} \tag{3.4}$$

Remark 2. We say that a sequence of numbers $d_0, \ldots, d_n$ is *symmetric* if for every whole number k from 0 to n,

$$d_k = d_{n-k}. \tag{3.5}$$

For example, the four-element sequence 1, 0, 0, 1 is symmetric.

A sequence $s_0, \ldots, s_{n+1}$ which is derived from a symmetric sequence $d_0, \ldots, d_n$ by Pascal's relations is itself symmetric. To establish this, we must verify the relations

$$s_k = s_{(n+1)-k} \tag{3.6}$$

for $k = 0, 1, 2, \ldots, n + 1$. But for $k = 0$ and $k = n + 1$, the equation (3.6) follows from the relations (3.1) and (3.3) and the condition $d_0 = d_n$ (which we get from (3.5) for $k = 0$). For $1 \leq k \leq n$, we have:

$$\begin{aligned} s_k &= d_{k-1} + d_k = d_{n-(k-1)} + d_{n-k} = d_{(n+1)-k} + d_{[(n+1)-k]-1} \\ &= d_{[(n+1)-k]-1} + d_{(n+1)-k} = s_{(n+1)-k}. \end{aligned} \tag{3.7}$$

In the case of our example, application of Pascal's relations to the sequence 1, 0, 0, 1 yields the five-element sequence 1, 1, 0, 1, 1 which is itself symmetric.

Let us now look at the sequence consisting of the single number 1. We shall call this sequence *Pascal's zeroth sequence*. From it, we may use Pascal's relations to generate a new sequence, which we shall call *Pascal's first sequence*. Applying Pascal's relations, we may then generate *Pascal's second sequence* from Pascal's first sequence, and so on. Since in each transition to a new sequence, the number of sequence elements is increased by one, there will be $n + 1$ numbers in Pascal's nth sequence. Without carrying out any calculations, we observe that remarks 1 and 2 allow us to conclude:

1. The sum of the numbers in Pascal's nth sequence is 2^n (since in proceeding from one sequence to the next, the sum of the numbers is doubled, and the sum of the numbers of the zeroth sequence is $2^0 = 1$).

2. All of Pascal's sequences are symmetric (since the property of symmetry is preserved in passing from one sequence to the next, and the zeroth sequence is symmetric).

Let us write down Pascal's sequences, one under another, so that each number of each sequence is found between and below those numbers of the preceding row from which it is calculated. We obtain an infinite table, called *Pascal's triangle*, which fills the interior of an angle; any

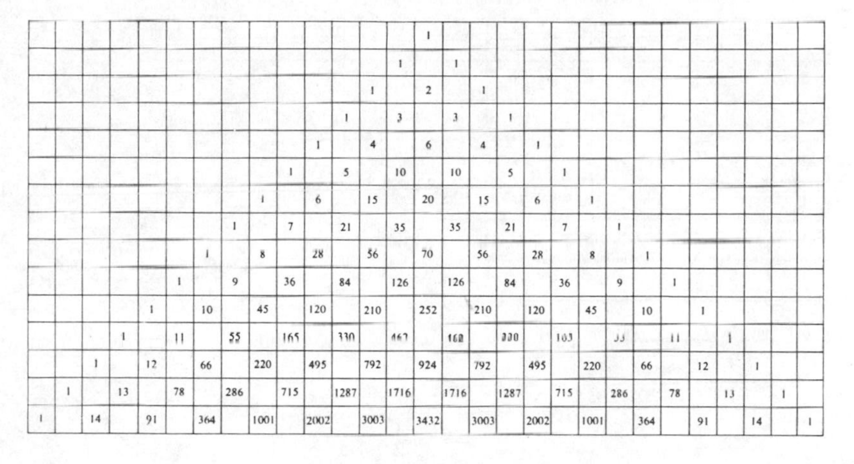

Fig. 3.1

segment of it, composed of the zeroth through *n*th rows, forms a triangle. A segment of Pascal's triangle, consisting of the first fifteen rows (the zeroth through the fourteenth) is presented in figure 3.1.

Pascal's triangle is symmetric about the bisector of the angle whose interior it fills, a consequence of the fact that each of its rows is symmetric. The numbers in it also satisfy a number of interesting properties. For example, the sum of the squares of the elements of any row is equal to some element of the vertical column along the bisector of the angle. For any prime number p, all elements of the pth row, except the first and last, are divisible by p.[1] (A *prime number* is a positive whole number whose only positive whole number divisors are itself and 1.)

It is clear that a method of constructing Pascal's triangle may be given without relying on the notions of "Pascal's relations" or "Pascal's sequences": Pascal's triangle is simply an infinite numerical table in "triangular form" in which each entry along the sides is 1, and in which each of the other entries is the sum of the two entries above it (to the right and to the left). The triangle first appeared in Pascal's paper "Treatise on the Arithmetic Triangle," published posthumously in 1665. In that work the table reproduced in figure 3.2 was published, in which each entry is the sum of the preceding entry in the same horizontal row and the preceding entry in the same vertical column.[2]

1. More of these properties are described on pp. 36–40 and 50–53 of the book *Problems in the Theory of Numbers* by E. B. Dynkin and V. A. Uspenskii (Boston: D. C. Heath and Company, 1963).
2. See B. Pascal, *Oeuvres completes*, vol. 3 (Paris: Hachette et Cie, 1908), p. 244.

1	1	1	1	1	1	1	1	1	1
1	2	3	4	5	6	7	8	9	
1	3	6	10	15	21	28	36		
1	4	10	20	35	56	84			
1	5	15	35	70	126				
1	6	21	56	126					
1	7	28	84						
1	8	36							
1	9								
1									

Fig. 3.2

Thus, what we call "Pascal's triangle" differs from the "triangle" examined by Pascal himself by a rotation through 45°.

Pascal investigated in detail the properties and applications of his "triangle"; several such applications will be examined in the following section. For the present, we shall examine three of the "triangle's" properties which were first noted by Pascal himself. For this purpose (and only at this point in our exposition) we shall consider that arrangement of the triangle in the plane which Pascal employed, and we shall speak of "rows" and "columns."

Property 1. Each number A in the table is the sum of the numbers in the preceding row, from the leftmost to the number which stands directly above A (see fig. 3.3, in which the squares containing the summands which give the sum A are shaded).

Property 2. Each number A in the table is the sum of the numbers in the preceding column, from the topmost to the number standing directly to the left of A (fig. 3.4).

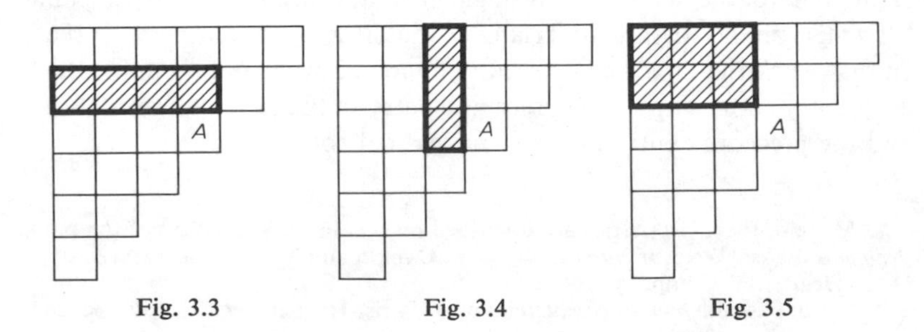

Fig. 3.3 Fig. 3.4 Fig. 3.5

Property 3. For each number A in the table, $A - 1$ is equal to the sum of all the numbers contained in the rectangle bounded by that column and that row whose intersection is the entry A (this row and column is not included in the rectangle; see fig. 3.5).

The proof of property 1 is by *mathematical induction*, a convenient method of proof for assertions about the nonnegative integers (whole numbers). The proof of such an assertion for all nonnegative integers m involves two steps: (1) establishing the assertion for $m = 0$; (2) a proof that the validity of the assertion for $m = k$ implies its validity for $m = k + 1$.

Once these two requirements are satisfied, the desired assertion is proved for all nonnegative integers m, for since the truth of the assertion for $m = 0$ is established in the first step, the second step allows us to conclude that the assertion is true for $m = 1$. Applying the second step again, we may conclude that the assertion is true for $m = 2$, and so on.

Property 1 may be proved by mathematical induction as follows: Number the rows and columns (starting from the top and left) of the triangle pictured, starting with zero. Let $A^n{}_m$ denote the entry in the nth row and mth column. The assertion is that $A^n{}_m$ is the sum of the first $m + 1$ entries of the $(n - 1)$th row, or

$$A^n{}_m = A^{n-1}{}_0 + A^{n-1}{}_1 + \cdots + A^{n-1}{}_m . \qquad (3.8)$$

Step 1. If $m = 0$, equation (3.8) becomes

$$A^n{}_0 = A^{n-1}{}_0 .$$

Since $A^n{}_0 = A^{n-1}{}_0 = 1$, the assertion holds for $m = 0$.

Step 2. Assuming (3.8) for $m = k$, and using the fact that each "interior" entry is the sum of the entry immediately preceding it in its column and the entry immediately preceding it in its row, we obtain

$$\begin{aligned} A^n{}_{k+1} &= A^n{}_k + A^{n-1}{}_{k+1} \\ &= A^{n-1}{}_0 + A^{n-1}{}_1 + \cdots + A^{n-1}{}_k + A^{n-1}{}_{k+1} \qquad (3.9) \end{aligned}$$

by using our assumption ("inductive hypothesis") on k, that

$$A^n{}_k = A^{n-1}{}_0 + A^{n-1}{}_1 + \cdots + A^{n-1}{}_k .$$

Since (3.9) is a restatement of (3.8) for $m = k + 1$, we have shown that the truth of our assertion (property 1) for $m = k$ implies its truth for $m = k + 1$. This completes the second step of the proof.

Property 2 may be proved by performing an induction on n rather than m; property 3 follows either from property 1 by induction on n or from property 2 by induction on m.

More than a century before Pascal's treatise, however, an interesting table—not in "triangular" but in "rectangular" form—was published in the *General Treatise on Number and Measure* (published in 1556–60), which also appeared after the death of its author, the distinguished Italian mathematician Niccolo Tartaglia (1506–59). Tartaglia's table had the form shown in figure 3.6.[3]

1	1	1	1	1	1
1	2	3	4	5	6
1	3	6	10	15	21
1	4	10	20	35	56
1	5	15	35	70	126
1	6	21	56	126	252
1	7	28	84	210	462
1	8	36	120	330	792

Fig. 3.6

Here each entry in the zeroth row is 1; in each of the remaining rows the leftmost (zeroth) entry is 1, and each succeeding entry is formed as the sum of the two entries directly before it and above it. The table which Tartaglia introduced is called (naturally) "Tartaglia's rectangle."

The elements of each of Pascal's sequences are usually numbered from left to right, beginning with the zeroth. Thus, the second place in the fifth row is occupied by the number 10. The number occupying the kth place in the nth row will be denoted by ${T^n}_k$ so that, for example, ${T^0}_0 = 1$, ${T^5}_2 = 10$, ${T^{14}}_4 = 1001$. The expression ${T^n}_k$ will obviously be defined for any $n \geq 0$ and $k = 0, 1, \ldots, n$.

Let us examine the infinite sequence formed by the numbers ${T^n}_k$ for any fixed k and variable n, that is, the sequence

$${T^k}_k, {T^{k+1}}_k, {T^{k+2}}_k, \ldots, {T^n}_k, \ldots. \tag{3.10}$$

This sequence begins with ${T^k}_k$ since the kth row is the first row which has a kth entry. Its elements are the numbers in Pascal's triangle occurring in the "kth line from the left, parallel to the left side," and also, because of the triangle's symmetry, the numbers occurring in the "kth

3. See G. G. Tseiten, *Istoriya matematiki v XVI i XVII vekakh* [The history of mathematics in the sixteenth and seventeenth centuries] (Moscow-Leningrad: GONTI, 1938), p. 116.

line from the right, parallel to the right side." In Tartaglia's rectangle, these numbers fill the kth column and the kth row.

For $k = 0$ we get the sequence

$$1, 1, 1, 1, 1, 1, \ldots$$

(the zeroth column or the zeroth row in Tartaglia's rectangle).

For $k = 1$ we get the sequence of natural numbers

$$1, 2, 3, 4, 5, 6, \ldots$$

(the first row or first column of Tartaglia's rectangle).

For $k = 2$ we get the sequence

$$1, 3, 6, 10, 15, 21, \ldots$$

(the second row or second column of Tartaglia's rectangle). The elements of this sequence are called *triangular numbers*; the mth triangular number is $T^{m+1}{}_2$, so that 1 is the first triangular number, 3 is the second triangular number, and so on. This name is a result of the fact that the mth triangular number $T^{m+1}{}_2$ is the number of spheres (or other identical objects) which can be packed in the shape of an equilateral triangle whose base is made up of m spheres (see fig. 3.7). In particular, the mth triangular number is the number of elements contained in the first m rows of Pascal's triangle, from the zeroth to the $(m - 1)$th.

Letting $k = 3$, we get the sequence

$$1, 4, 10, 20, 35, 56, \ldots$$

(the third row or third column of Tartaglia's rectangle). The numbers of this sequence are called *pyramidal numbers*, or more precisely, *tetrahedral numbers*; 1 is the first tetrahedral number, 4 the second, 10 the

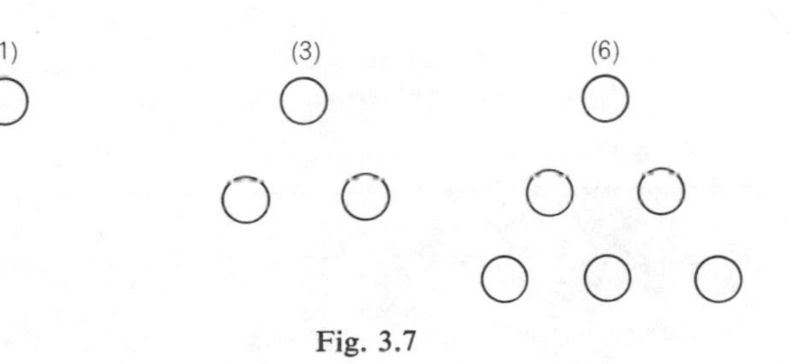

Fig. 3.7

third, and so on, so that the mth tetrahedral number is $T^{m+2}{}_3$. The mth tetrahedral number $T^{m+2}{}_3$ is the number of spheres which can be packed in the shape of a tetrahedron (triangular pyramid) with an equilateral triangular base of side m (see fig. 3.8).[4]

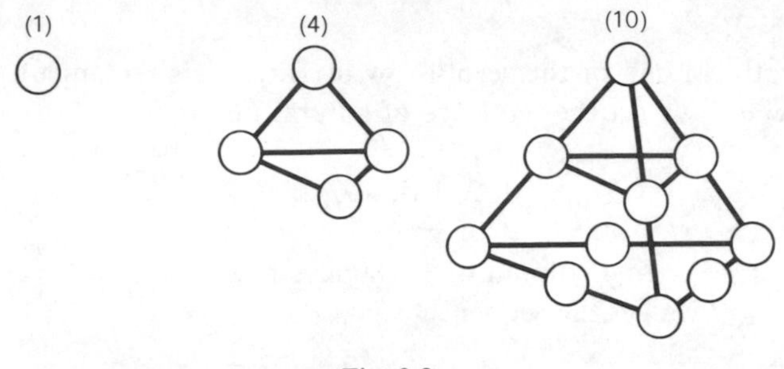

Fig. 3.8

4. Triangular and pyramidal numbers (which are special cases of the so-called figure numbers) were of interest to the ancient Greeks, who attributed mystical properties to them. Of the writings preserved today in which these numbers are examined, the earliest is probably *Introduction to Arithmetic*, by the ancient Greek mathematician Nicomachus of Gerasa, who lived around A.D. 100. See D. J. Struik, *A Concise History of Mathematics* (New York: Dover Publications, 1967), p. 72; B. L. van der Waerden, *Science Awakening* (New York: Oxford University Press, 1961, pp. 98–100). According to indirect information, however, *polygonal numbers* were studied considerably earlier, in the 2d century B.C., and even earlier, in the 5th century B.C. by the famous mathematician Pythagoras and his students, the Pythagoreans (see pp. 46–47 of the above book by D. J. Struik).

4 Pascal's Operation

By virtue of their definition, the numbers $T^n{}_k$ are subject to the following relations:

$$T^0{}_0 = 1, \qquad (4.1)$$

$$T^{n+1}{}_0 = T^{n+1}{}_{n+1} = 1 \quad \text{for } n = 0, 1, 2, \ldots, \qquad (4.2)$$

$$T^{n+1}{}_k = T^n{}_{k-1} + T^n{}_k \quad \text{for } n = 0, 1, 2, \ldots; k = 1, 2, \ldots, n. \qquad (4.3)$$

The numbers $T^n{}_k$ are completely determined by these relations; using the equations (4.1)–(4.3), we may construct as many rows of Pascal's triangle as we wish.

The definition of $T^n{}_k$ may be extended in a natural way so that it makes sense for any nonnegative integer n and any integer k. To do this, we set $T^n{}_k = 0$ for $n \geq 0$ and for k such that $0 > k$ or $k > n$. Thus, $T^n{}_k = 0$ for all pairs (n, k) for which $n \geq 0$, $k < 0$, and for all pairs (n, k) for which $n \geq 0$, $k > n$. The relation $T^{n+1}{}_k = T^n{}_{k-1} + T^n{}_k$ will then be satisfied for all k (and not only for k from 1 to n, as in [4.3]), and the numbers $T^n{}_k$ will be completely determined by the following equations:

$$T^0{}_0 = 1, \qquad (4.4)$$

$$T^0{}_k = 0 \qquad \text{for } k \neq 0, \qquad (4.5)$$

$$T^{n+1}{}_k = T^n{}_{k-1} + T^n{}_k \quad \text{for all } n \geq 0 \text{ and all } k. \qquad (4.6)$$

These relations permit us to give a graphic representation of the generation of Pascal's triangle. Let us consider an infinite table of zeroes, arranged in staggered rows, as shown below:

```
 .... 0 0 0 0 ....
.... 0 0 0 0 0 ....
 .... 0 0 0 0 ....
.... 0 0 0 0 0 ....
..................
```

It is clear that such a table satisfies Pascal's relations, which require each number to be the sum of the two nearest numbers in the preceding row.[1] We now imagine that one of the zeroes in the first row of this table is replaced by a one. If Pascal's relations are to be preserved, then the "perturbation" will "enlarge to an angle"—just like a wave in a brook when disturbed by a stick—in the form of Pascal's triangle:

```
 .... 0 0 1 0 0 ....
.... 0 0 1 1 0 0 ....
 .... 0 1 2 1 0 ....
.... 0 1 3 3 1 0 ....
.....................
```

Given arbitrary n and k ($n = 0, 1, 2, \ldots; k = 0, 1, \ldots, n$), it would be possible to find T^n_k, if we had sufficient time and patience, by writing out Pascal's triangle and continuing until we arrived at the kth number of the nth row. Or we could take advantage of the relations (4.1)–(4.3) which permit us to determine T^n_k by performing a finite number of additions.

We leave it as a problem for the reader to determine the *minimum* number of additions which must be carried out to calculate T^n_k, using the relations (4.1)–(4.3), for given n and k. (Hint: Try to take advantage of the symmetry of Pascal's triangle.)

1. We will consider an infinite row

$$\ldots, s_{-3}, s_{-2}, s_{-1}, s_0, s_1, s_2, \ldots$$

to be derived from the infinite row

$$\ldots, d_{-2}, d_{-1}, d_0, d_1, d_2, \ldots$$

by Pascal's law, when $s_k = d_{k-1} + d_k$ for each k. The definition of Pascal's law for finite rows follows from this if each finite row

$$x_0, x_1, \ldots, x_n$$

is identified with the infinite row

$$\ldots, 0, 0, 0, x_0, x_1, \ldots, x_n, 0, 0, 0, \ldots.$$

Let us agree to call the operation of calculating T^n_k from the numbers k and n, *Pascal's operation*. Pascal's operation is then defined for any n and k for which $n \geq 0, 0 \leq k \leq n$.[2]

But if we redefine T^n_k according to relations (4.4)–(4.6), then Pascal's operation will be defined for any nonnegative integer n and any integer k.

With the help of Pascal's operation, it is easy to write down the numbers H^n_k, which serve as a solution to the Olympiad problem of section 1. To find these numbers, we first define (for $m = 0, 1, \ldots, 1000$; $q = 0, 1, \ldots, m$)

$$Z^m_q = \frac{1}{2^{1000-m}} H^m_q \tag{4.7}$$[3]

so that

$$H^m_q = 2^{1000-m} Z^m_q. \tag{4.8}$$

Then, from the relations (4.7) and (1.1), we get

$$Z^0_0 = \frac{1}{2^{1000}} \cdot 2^{1000} = 1. \tag{4.9}$$

In the relations (1.2), (1.4), and (1.3) we may then replace the number H^m_q by its expression in terms of Z^m_q given in (4.8). We get, from (1.2),

$$2^{1000-(n+1)} Z^{n+1}_0 = \frac{2^{1000-n} Z^n_0}{2},$$

from which we obtain

$$Z^{n+1}_0 = Z^n_0. \tag{4.10}$$

In exactly the same way, from (1.4) we get

$$2^{1000-(n+1)} Z^{n+1}_{n+1} = \frac{2^{1000-n} Z^n_n}{2},$$

2. Pascal himself (proceeding from the rectangular arrangement of the table, which he proposed—see fig. 3.2, p. 12, above) examined a different operation in his treatise, that of finding the number standing at the intersection of the xth column and the yth row (with the rows and columns numbered beginning with the first, so that this operation is defined for $x \geq 1, y \geq 1$) from the numbers x and y. If this number is denoted by $P(x, y)$, then, as one may easily verify, $P(x, y) = T^{x-1+y-1}_{x-1}$, from which $T^n_k = P(k + 1, n - k + 1)$.

3. Since 2 to the zeroth power is considered to be equal to 1, for $m = 1000$, equation (4.7) takes the form $Z^{1000}_q = H^{1000}_q$.

from which

$$Z^{n+1}{}_{n+1} = Z^n{}_n\,. \tag{4.11}$$

Finally, from (1.3) we get

$$2^{1000-(n+1)}Z^{n+1}{}_k = \frac{2^{1000-n}.Z^n{}_{k-1} + 2^{1000-n}.Z^n{}_k}{2}\,,\ 1 \le k \le n,$$

from which

$$Z^{n+1}{}_k = Z^n{}_{k-1} + Z^n{}_k\,,\ 1 \le k \le n. \tag{4.12}$$

The equations (4.10)–(4.12) show that each sequence

$$\omega_{n+1} = \langle Z^{n+1}{}_0, \ldots, Z^{n+1}{}_{n+1}\rangle\,,$$

where $n = 0, 1, \ldots, 999$ is obtained from the preceding sequence

$$\omega_n = \langle Z^n{}_0, \ldots, Z^n{}_n\rangle$$

according to Pascal's relations. Since, as is clear from equation (4.9), the initial sequence

$$\omega_0 = \langle Z^0{}_0\rangle$$

is Pascal's zeroth sequence, then the sequence following it, ω_1, is Pascal's first sequence; the sequence ω_2 is Pascal's second sequence; and so on. For each m from 0 to 1000,[4] the sequence ω_m is Pascal's mth sequence, and

$$Z^m{}_q = T^m{}_q\,. \tag{4.13}$$

Consequently, by (4.8), for each $m = 0, 1, \ldots, 1000$ and for each $q = 0, 1, \ldots, m$,

$$H^m{}_q = 2^{1000-m}T^m{}_q\,. \tag{4.14}$$

In particular, for $m = 1000$,

$$H^{1000}{}_q = T^{1000}{}_q\,. \tag{4.15}$$

4. For $m > 1000$, the sequence ω_m is not defined.

Thus, the number of people who arrive at an intersection in the thousandth row is simply an element of Pascal's thousandth sequence! If Pascal's operation is considered to be standard, then equation (4.15) gives a solution to the problem of section 1 (in the third form discussed at the end of chapter 2). In the following two chapters we shall see how two important problems can be solved with the aid of Pascal's operation.

5 Binomial Coefficients

In this section we shall find expressions for the so-called binomial coefficients by using Pascal's operation. In order to define the binomial coefficients, we take the binomial $1 + x$ and raise it to powers 0, 1, 2, 3, . . . , arranging the terms of the resulting polynomials in order of ascending powers of the symbol x. We get

$$(1 + x)^0 = 1, \quad (5.1)$$

$$(1 + x)^1 = 1 + x, \quad (5.2)$$

$$(1 + x)^2 = (1 + x)(1 + x) = 1 + 2x + x^2, \quad (5.3)$$

$$(1 + x)^3 = (1 + x)^2(1 + x) = 1 + 3x + 3x^2 + x^3, \quad (5.4)$$

and so on.

In general, for any nonnegative integer n,

$$(1 + x)^n = a_0 + a_1x + a_2x^2 + \cdots + a_px^p, \quad (5.5)$$

where $a_0, a_1, \ldots, a_p$ are constants. If you wish, you can easily verify that $p = n$ and that $a_0 = a_p = 1$; however, we do not need this now. Somewhat later on, we will obtain this result as a consequence of a more general formula. At this stage, it is sufficient for us to know that the result of raising the binomial $1 + x$ to the power n (where n is a nonnegative integer) may be written as a polynomial with integral coefficients, arranged in order of increasing powers of the letter x, as exhibited in the relation (5.5). This polynomial is called the *binomial expansion of* $(1 + x)^n$. Of course, its coefficients (and $p + 1$, the number of them) depend on n. In order to stress this dependence, one often makes use of expressions for these coefficients in which n appears. Specifically, the coefficient of x^k in the binomial expansion of $(1 + x)^n$ will be designated by $\binom{n}{k}$. The numbers $\binom{n}{k}$ are called *binomial coefficients.*

The relation (5.5) may now be written as

$$(1 + x)^n = \binom{n}{0} + \binom{n}{1}x + \binom{n}{2}x^2 + \cdots + \binom{n}{p}x^p, \tag{5.6}$$

and from the relations (5.1)–(5.4), we get

$$\binom{0}{0} = 1$$

$$\binom{1}{0} = 1 \quad \binom{1}{1} = 1$$

$$\binom{2}{0} = 1 \quad \binom{2}{1} = 2 \quad \binom{2}{2} = 1$$

$$\binom{3}{0} = 1 \quad \binom{3}{1} = 3 \quad \binom{3}{2} = 3 \quad \binom{3}{3} = 1.$$

We see that for the exponents $n = 0, 1, 2, 3$, the rows of binomial coefficients coincide respectively with the 0th, 1st, 2d, and 3d rows of Pascal's triangle. We shall now show that the analogous relations hold for each n. To do this, we shall look at how the sequence of coefficients for $(x + 1)^{n+1}$ is derived from the sequence of coefficients for $(x + 1)^n$, taking advantage of the formula

$$(1 + x)^{n+1} = (1 + x)^n(1 + x). \tag{5.7}$$

Let us write down the expansions for the left and right sides of this formula in ascending powers of the letter x. For the left side, formula (5.6) gives (by substituting $n + 1$ for n)

$$(1 + x)^{n+1} = \binom{n+1}{0} + \binom{n+1}{1}x + \cdots + \binom{n+1}{k}x^k + \cdots + \binom{n+1}{q}x^q, \tag{5.8}$$

for some q. By virtue of the same formula (5.6), we have for the right side:

$$(1 + x)^n(1 + x)$$

$$= \left[\binom{n}{0} + \binom{n}{1}x + \cdots + \binom{n}{p}x^p\right](1 + x)$$

$$= \binom{n}{0} + \binom{n}{1}x + \cdots + \binom{n}{k}x^k + \cdots + \binom{n}{p}x^p$$

$$+ \binom{n}{0}x + \cdots + \binom{n}{k-1}x^k + \cdots + \binom{n}{p-1}x^p + \binom{n}{p}x^{p+1}$$

$$= \binom{n}{0} + \left[\binom{n}{0} + \binom{n}{1}\right]x + \cdots + \left[\binom{n}{k-1} + \binom{n}{k}\right]x^k + \cdots$$

$$+ \left[\binom{n}{p-1} + \binom{n}{p}\right]x^p + \binom{n}{p}x^{p+1}. \tag{5.9}$$

Because of (5.7), the right sides of (5.8) and (5.9) are equal. Therefore, $q = p + 1$; equating coefficients for identical powers of the letter x, we get

$$\binom{n+1}{0} = \binom{n}{0}, \tag{5.10}$$

$$\binom{n+1}{k} = \binom{n}{k-1} + \binom{n}{k} \quad \text{if } 0 < k < p + 1, \tag{5.11}$$

$$\binom{n+1}{p+1} = \binom{n}{p}. \tag{5.12}$$

The relations (5.10)–(5.12) show that the sequence of coefficients of the binomial expansion of $(x + 1)^{n+1}$ are derived from the sequence of coefficients of the binomial expansion of $(x + 1)^n$ by Pascal's law. Since the sequence of coefficients of the binomial expansion of $(x + 1)^0$ coincides with Pascal's zeroth sequence, all succeeding sequences of coefficients also coincide with the corresponding rows of Pascal's triangle. Therefore, the numbers $\binom{n}{k}$ are defined only for $k = 0, 1, \ldots, n$, with

$$\binom{n}{k} = T^n{}_k. \tag{5.13}$$

Remark. The question "What coefficients do x^{-3} and x^{20} have in the binomial expansion of $(x + 1)^5$?" may be answered, "The coefficients are zero." Therefore, the expression $\binom{n}{k}$ may be defined in a natural manner for the cases $k < 0$ and $k > n$ by setting $\binom{n}{k} = 0$ in these cases. Then the equation (5.13) will hold true for all nonnegative n and all integers k, by virtue of the redefinition of the symbol $T^n{}_k$ which was made in the preceding chapter.

Thus, we have expressed the binomial coefficients in terms of Pascal's operation. We may now rewrite equation (5.6) in the following form:

$$(1 + x)^n = T^n{}_0 + T^n{}_1 x + T^n{}_2 x^2 + \cdots + T^n{}_k x^k + \cdots + T^n{}_n x^n. \quad (5.14)$$

Formula (5.14) is sometimes called *Newton's binomial formula*, or simply *Newton's formula*.[1] Another more traditional expression of this formula will be presented in chapter 7.

In a certain sense, this section has provided a "solution" to the problem of finding an expression for the binomial coefficient $\binom{n}{k}$. Recalling chapter 2, we know that we have more than one criterion for the "solution" of a problem. For example, if (under the second interpretation) a solution is considered to be an expression which allows us to get the binomial coefficient $\binom{n}{k}$ from n and k, then $\binom{n}{k}$ is itself a solution. If we require that the solution express $\binom{n}{k}$ in terms of the numbers n and k and certain standard operations (as in the third interpretation), our concept of "solution" will depend on the collection of standard operations chosen. If Pascal's operation is considered standard, then (5.13) is a solution to the problem of finding the binomial coefficients $\binom{n}{k}$. Another solution to this problem, corresponding to a different collection of standard operations, will be given in chapter 7.

1. Formula (5.14) was known long before Newton; in particular, Tartaglia had already mentioned it. Newton's name is connected with the formula only because he pointed out a method of generalizing this formula to the case of an arbitrary rational (including negative) exponent in 1676.

6 The Number of Subsets of a Given Set

In mathematics, any collection of objects is called a *set*. Thus,

(a) the collection of all pages in this booklet,
(b) the collection of all integers,
(c) the collection of all even numbers,
(d) the collection of all the pencils in a certain box

are all sets.

If some object and some set are given, exactly one of the following two statements is true:

1. The object belongs to the set.
2. The object does not belong to the set.

In the first case, the object is called an *element* of the set. For example, the number 3 is an element of the set of all integers, and is not an element of the set of all even numbers.

It may happen that all elements of some set A are elements of another set B (for instance, all elements of the set of all even numbers are elements of the set of all integers). In such a case, the set A is said to be *contained* in, or a *subset* of, the set B. Obviously, every set is a subset of itself. If the set A is a subset of the set B, and the set B is a subset of the set A, then A and B consist of the very same elements, and are *equal*.

Sets can be *finite* (like the sets in examples (a) and (d) above), or *infinite* (like the sets in examples (b) and (c) above). Finite sets (and we will study only such sets in this section) are the subject of a particular discipline in mathematics—that of *combinatorial analysis*.

A particular set stands out among finite sets: the set containing no elements, the so-called *empty set*. Thus, the possibility is taken into account that on opening the box in example (d), we discover that the set of all pencils contained in it is empty. The empty set is considered to be a subset of every set.

If a set is finite, then its elements may be numbered to find how many elements are in the set. A set which consists of n elements is called an *n element* set. The set of pages of this book is a 44-element set, for example, and the empty set is a zero-element set.

Example. Let us examine a set consisting of three objects, a pencil, a pen, and an eraser, and determine all of its subsets. There is exactly one zero-element subset, the empty set. There are exactly three one-element subsets (fig. 6.1).

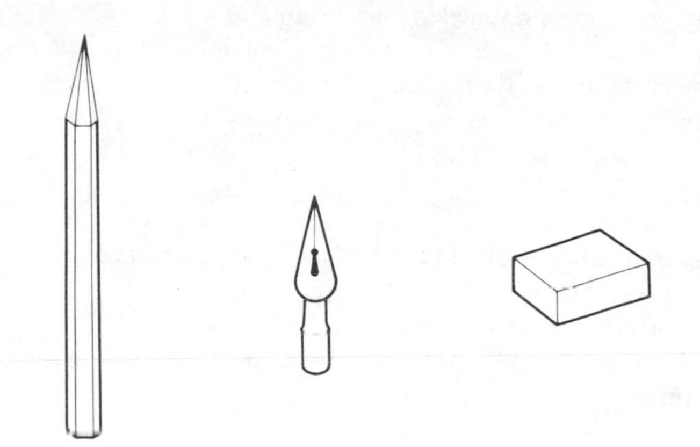

Fig. 6.1

There are exactly three two-element subsets (fig. 6.2).

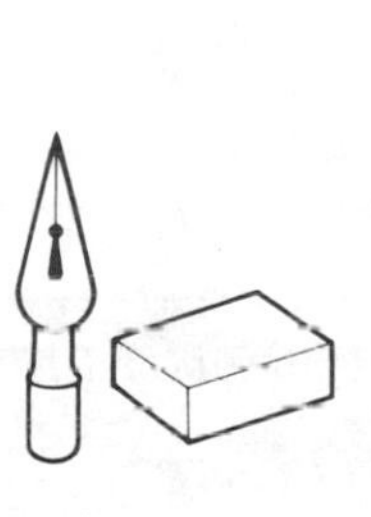

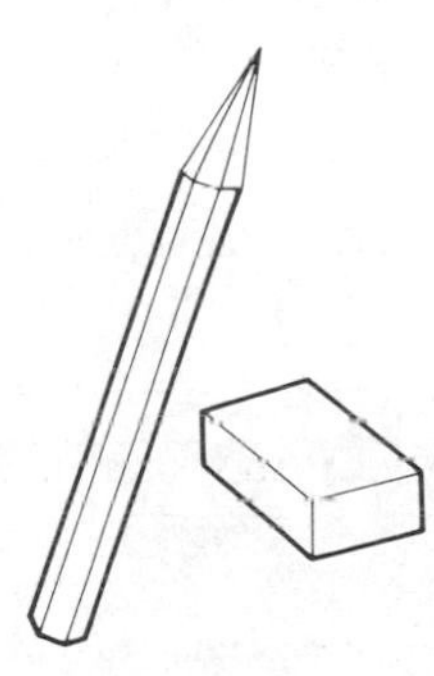

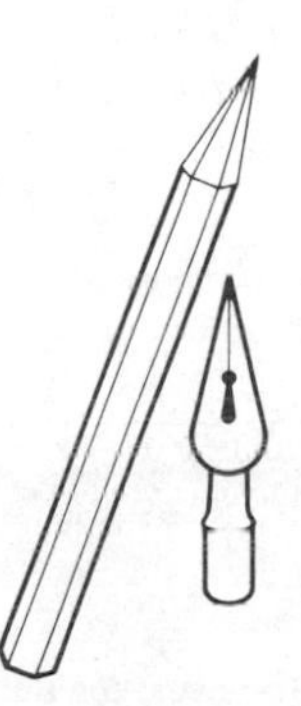

Fig. 6.2

Finally, there is exactly one three-element subset (the set itself) (fig. 6.3).

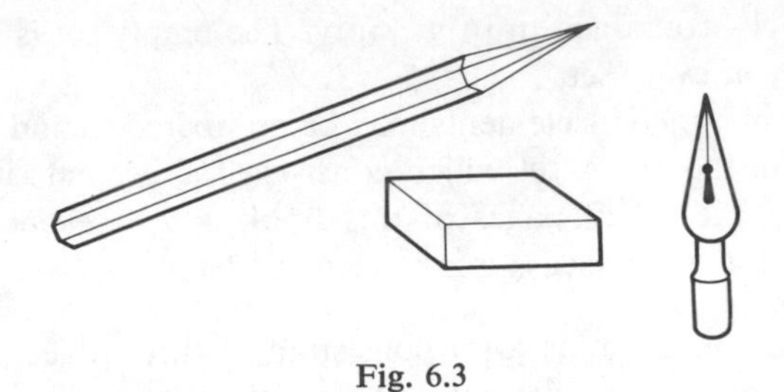

Fig. 6.3

Thus, our set has eight subsets in all.

Let an n-element set be given; any k-element subset of it is called a *combination* of the n given elements taken k at a time. It is obvious that the number of combinations of n given elements taken k at a time does not depend on the n given elements, but only on the numbers n and k. The *number of combinations of n elements taken k at a time* is denoted

$$C^n_k .$$

Put differently, C^n_k is the number of k-element subsets of an n-element set. The expression C^n_k is usually considered to make sense for $n = 0, 1, 2, \ldots$, and $0 \leq k \leq n$.[1]

The total number of subsets of an n-element set will be denoted by C_n, so that

$$C_n = C^n_0 + C^n_1 + \cdots + C^n_n . \tag{6.1}$$

What are the numbers C_n and C^n_k? We can answer this question in a few specific cases at once. From the example just investigated, we know that $C_3 = 8$, $C^3_0 = C^3_3 = 1$, and $C^3_1 = C^3_2 = 3$.

Furthermore, we can verify the three properties listed below.

First property of the number of combinations:

$$C^m_0 = C^m_m = 1 . \tag{6.2}$$

Proof. It is clear that any m-element set S has exactly one zero-element subset (the empty set) and one m-element subset (the set S itself).

1. However, the definition can be extended to make sense for $k > n$ by setting it equal to zero in this case (since for $k > n$ no k-element subset exists).

Without actually calculating the numbers C^n_k, we shall now establish two more properties of these numbers. The proof of the second property is a helpful exercise in the mastery of the concepts put forth in this section; and the third property, together with the first, is a basis for calculating the numbers C^n_k.

Second property of the number of combinations:

$$C^n_k = C^n_{n-k}\,. \tag{6.3}$$

Proof. Let us consider any n-element set M. We must show that the number of k-element subsets of M is equal to the number of $(n-k)$-element subsets of M. Let us carry out the following construction mentally. From paper, we cut out as many squares as we have k-element subsets of our set (that is, C^n_k) and on each of them, we write out one of these subsets, so that each k-element subset will be listed on exactly one square. Let us also cut from paper C^n_{n-k} circles, writing each $(n-k)$-element subset on some circle. It is now sufficient for us to show that there are equal numbers of circles and squares. For this purpose, we lay all the squares on a table, and on each of them we place a circle, according to the following rule: If some k-element subset of the set M is listed on a square, we place on this square the circle on which is listed the subset of the set M consisting of the remaining elements

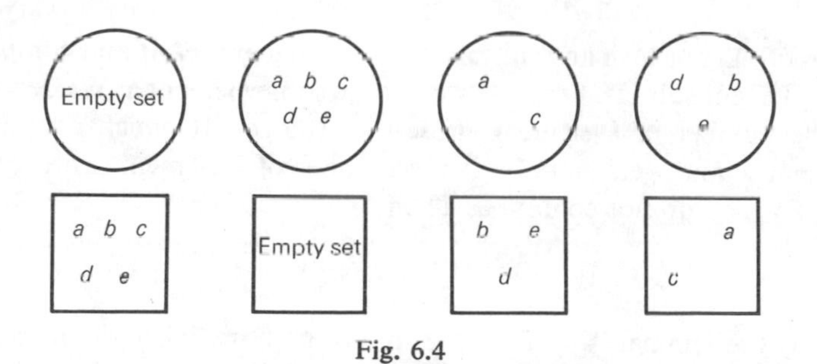

Fig. 6.4

(for the case of a five-element set M, consisting of the elements a, b, c, d, e, several squares together with their corresponding circles are shown in figure. 6.4). It is evident that on each square there lies precisely one circle, and that each circle will be placed on precisely one square, implying that there are exactly as many circles as squares.

Before going on to the third property, let us prove the following lemma.

LEMMA. *Let us choose some element a in an* $(n + 1)$*-element set S. The number of k-element subsets of this set which contain this chosen element is equal to* $C^n{}_{k-1}$.

Proof. Again, let us conduct a mental experiment with circles and squares. We cut from paper as many squares as there are k-element subsets containing the chosen element, and on each of these we list one such subset, so that each of them will be represented once. We then cut from paper $C^n{}_{k-1}$ circles, and on each circle we list one of the $(k - 1)$-element subsets of the n-element set of all unchosen elements, so that all such subsets will be depicted (there are n unchosen elements, and therefore $C^n{}_{k-1}$ such subsets). On each square we place a circle, according to the following rule: If some subset A is depicted on a square, then on that square must be placed the circle listing the set derived from A by removing the chosen element. It is clear that on each square there lies exactly one circle, and that each circle is placed on exactly one square, implying that the number of squares and the number of circles are both equal to $C^n{}_{k-1}$. Since we cut out as many squares as there are k-element subsets of the $(n + 1)$-element set containing the chosen element, the number of such subsets is equal to $C^n{}_{k-1}$, which is what we were required to prove.

We now go on to the third property of the number $C^n{}_k$.

Third property of the number of combinations:

$$C^{n+1}{}_k = C^n{}_{k-1} + C^n{}_k\,, \; 1 \leq k \leq n. \tag{6.4}$$

Proof. Let us take an arbitrary $(n + 1)$-element set M and compile all its k-element subsets. From M we choose some element a. We denote by X the number of k-element subsets of the set M which contain the element a, and we denote by Y the number of k-element subsets of the set M which do not contain a. Then

$$C^{n+1}{}_k = X + Y. \tag{6.5}$$

But by the lemma, $X = C^n{}_{k-1}$. Moreover, Y is simply the number of combinations of the n unchosen elements taken k at a time, that is, $C^n{}_k$. Therefore,

$$C^{n+1}{}_k = C^n{}_{k-1} + C^n{}_k\,, \tag{6.6}$$

which is what we wanted to show.

The third property, together with the first, shows that the sequence

$$C^{n+1}{}_0, C^{n+1}{}_1, \ldots, C^{n+1}{}_{n+1} \tag{6.7}$$

is derived from the sequence

$$C^n{}_0, C^n{}_1, \ldots, C^n{}_n \tag{6.8}$$

by Pascal's relations. Since for $n = 0$ the sequence

$$C^0{}_0 \tag{6.9}$$

coincides with Pascal's zeroth sequence, we know that for any arbitrary n the sequence (6.8) will coincide with Pascal's nth sequence, and thus,

$$C^n{}_k = T^n{}_k . \tag{6.10}$$

Thus, we have a means of calculating the number of k-element subsets of an n-element set, that is, the number of combinations of n elements taken k at a time (in this way, formula [6.10] gives a solution to the "problem of the number of combinations," under the condition that Pascal's operation is considered standard).[2]

Finally, the relations (6.1) and (6.10) show that the number of all subsets of an n-element set is equal to the sum of all the entries in Pascal's nth sequence. As we know, this sum is equal to 2^n. Consequently,

$$C_n = 2^n . \tag{6.11}$$

2. The reader will find a different solution, with different standard operations, in section 7.

7 The Connection with Factorials

In chapter 4, two methods of calculating the number T^n_k from the numbers n and k were pointed out: the more "mechanical" method of writing out Pascal's triangle (which, however, leads to superfluous calculations), and the method more economical with regard to the number of steps (which, however, requires a certain organization of calculations), consisting of repeated application of the relations (4.1)–(4.3). These two methods are very similar, for in both cases the numbers T^n_k are obtained using Pascal's relations. However, there is another method of finding T^n_k, which we shall now discuss.

First, let us introduce a new symbol. We set

$$0! = 1,$$

and for each whole number m, we define

$$m! = (m-1)!\,m.$$

Thus, for $m > 0$,

$$m! = 1\cdot 2\cdot \cdots \cdot m.$$

The expression $m!$ is read "m factorial."

We shall now express Pascal's operation in terms of arithmetic operations and the operation of taking factorials. For this purpose, we examine the following expression:

$$\frac{m!}{q!\,(m-q)!}.$$

Let us denote this expression by $F^m{}_q$. It is clear that the expression $F^m{}_q$ makes sense for $m \geq 0$, $0 \leq q \leq m$. We notice that

$$F^0{}_0 = \frac{0!}{0!\,0!} = 1.$$

Furthermore,

$$F^m{}_0 = \frac{m!}{0!\,m!} = 1, \qquad F^m{}_m = \frac{m!}{m!\,0!} = 1.$$

Finally,

$$\begin{aligned} F^n{}_{k-1} + F^n{}_k &= \frac{n!}{(k-1)!\,(n-k+1)!} + \frac{n!}{k!\,(n-k)!} \\ &= \frac{n!}{(k-1)!\,(n-k)!\,(n-k+1)} + \frac{n!}{(k-1)!\,k(n-k)!} \\ &= \frac{n!}{(k-1)!\,(n-k)!} \cdot \left[\frac{1}{n-k+1} + \frac{1}{k}\right] \\ &= \frac{n!}{(k-1)!\,(n-k)!} \cdot \frac{n+1}{k(n-k+1)} = \frac{(n+1)!}{k!\,(n+1-k)!} \\ &= F^{n+1}{}_k, \ 1 \leq k \leq n. \end{aligned}$$

Thus, the sequence

$$F^0{}_0$$

is Pascal's zeroth sequence, while the $(n+1)$th sequence

$$F^{n+1}{}_0, F^{n+1}{}_1, \ldots, F^{n+1}{}_{n+1}$$

is derived from the nth sequence

$$F^n{}_0, F^n{}_1, \ldots, F^n{}_n$$

by Pascal's relations. Therefore, for any $m = 0, 1, 2, \ldots$, the sequence

$$F^m{}_0, F^m{}_1, \ldots, F^m{}_m$$

coincides with Pascal's mth sequence, and

$$F^m{}_q = T^m{}_q.$$

Hence,

$$T^m{}_q = \frac{m!}{q!\,(m-q)!}\,.$$

We have thus expressed Pascal's operation in terms of the operations of taking factorials, subtraction, multiplication, and division, in the sense that we have found an expression for $T^m{}_q$ containing only m, q, and the symbols for the indicated operations. This permits us to calculate $T^m{}_q$ directly, since we are able to calculate factorials, differences, products, and quotients.

Several interesting corollaries follow quickly from the formula for $T^m{}_q$ just calculated.

COROLLARY 1. Canceling $(m - q)!$ in the numerator and denominator of the expression for $T^m{}_q$, we get

$$\begin{aligned} T^m{}_q &= \frac{m!}{q!\,(m-q)!} \\ &= \frac{m(m-1)\cdots[m-(q-1)](m-q)!}{q!\,(m-q)!} \\ &= \frac{m(m-1)\cdots[m-(q-1)]}{q!} \\ &= \frac{m(m-1)\cdots[m-(q-1)]}{q\cdot(q-1)\cdot\ \ldots\ \cdot 1}\,. \end{aligned}$$

COROLLARY 2. Let $m \geq 1$, $m \geq q \geq 1$. The product of the q factors $m(m - 1)\cdots[m - (q - 1)]$ is always divisible by the product of the q factors $1\cdot 2\cdots q$.

Specifically, because of corollary 1, the ratio of these products is equal to $T^m{}_q$, a whole number.

COROLLARY 3. From the relation (4.15), we get

$$H^{1000}{}_q = \frac{1000!}{q!\,(1000-q)!}\,.$$

This is a new form of the solution to the problem of section 1.

COROLLARY 4. From the relation (5.13), we get

$$\binom{n}{k} = \frac{n!}{k!\,(n-k)!} = \frac{n(n-1)\cdots[n-(k-1)]}{1\cdot 2\cdot\ \ldots\ \cdot k}\,.$$

This is the traditional high-school expression for the binomial coefficient.

COROLLARY 5. From the relation (5.14) and corollary 1, we conclude:

$$(1 + x)^n = 1 + nx + \frac{n(n-1)}{1\cdot 2}x^2 + \cdots$$
$$+ \frac{n(n-1)\cdots[n-(k-1)]}{1\cdot 2\cdot \ldots \cdot k}x^k + \cdots + x^n.$$

This is the traditional high-school form of Newton's binomial formula.

COROLLARY 6. The relation (6.10) gives the traditional high-school formula for the number of combinations

$$C^n_k = \frac{n!}{k!\,(n-k)!} = \frac{n(n-1)\cdots[n-(k-1)]}{1\cdot 2\cdot \ldots \cdot k}.$$